Farmotherlands

First published 2025 by The Hedgehog Poetry Press,
5 Coppack House, Churchill Avenue, Clevedon. BS21 6QW
www.hedgehogpress.co.uk

Copyright © Julie Leoni 2025

The right of Julie Leoni to be identified as the author of this work has been asserted in accordance with the Copyright, Designs and Patents Act 1988. All rights reserved. No part of this publication may be reproduced, stored in or introduced into a retrieval system, or transmitted in any form, or by any means (electronic, mechanical, photocopying, recording or otherwise) without prior written permissions of the publisher. Any person who does any unauthorised act in relation to this publication may be liable for criminal prosecution and civil claims for damages.

ISBN: 978-1-916830-36-3

Farmotherlands
by

Julie Leoni

Contents

Dawn 9
Bottled 10
Hiraeth 12
Herstories 13
Definition 14
I am not Seamus Heaney 15
Dead things 16
Cattle 18
Scar 22
Between the lines 23
Whet 24
January in Waen Field After You Died 26
November Walk at Lake Vyrnwy 27
2001 28
Ecclesiastes 3:1-8 29
Now and then 30
Quietus 31
Rooting 32
Milk's Gotta Lotta Bottle 34
Tipping point 35
Shipwrecked 36
Foreshadowing 37
Liminal 38
Driving to work 40
Feral 41
Witch 42

Craft	43
Trespassers W	44
Field	46
Shifting baseline syndrome	47
Confluence	48
Where was Noah when they needed him?	49
Gardening	50
Harvest	52
Care	53
Too Late	54
You don't know what you've got til its gone	56
Road-kill	57
Man Versus Great Birnam Wood	58
Vulpes Vulpes	59
Mink	60
Hawk	62
Soar	63
Firefly	64
Breakfast	65
Ducks	66
Asunder	67
A summer evening	68
Present	69
Hope	70
Meditation with a barn owl at Treflach Regenerative Farm	71
Hares at 42 Acres Regenerative Farm	72

*For the land, my family, especially my mum Elin
and always Matty and Ben*

DAWN

Blackbird stitches bright
song through layers of quilted sleep
tugging me awake

BOTTLED

Come down to the cellar with me. Mind the steps, the hanging lime plaster, ignore the mouse skittering through rafters and laugh at the rodent-proof wire on the dusty stone floor. Look, over there, in the far-left corner where the light doesn't reach. Top shelf. Those jars.

Those jars she washed, marigold hands in scalding water. Jars baked clean. Tin lids she pushed apart, the flat, gold top circled in red rubber, sitting snug inside the rim she screwed tightly onto the cooled, full jars. But first, first, *don't rush,* our job, small fingers, to lay those tiny, fiddly, *make sure they don't get stuck together* waxy circles of paper over the hot boiled fruit. So sweet. Tempted, but sugar burned so we did not stick our fingers deep to lick.

Bottled fruit: rhubarb, raspberry, strawberry, damson, blackberry. Spring rhubarb from our garden, the patch at the back behind the apple tree where our garden touched the sherbet-lemon fence of Half-Pint and Danny.

Raspberries calling ruby. Forbidden-few, hedging behind the rhubarb. I would sneak and pick just one, just one, pop it from the white plug of plant to crown my little finger, then parting my lips, I would search for juice, furtively inserting my eager tongue.

A rare treat, to make just one jar, just one. Stir. Spoon one drop, *just one,* onto the cold Co-op saucer, then into the fridge until the test of eager fingers, wrinkling congealed summer-blood. *Is it ready? Is it ready? Can I lick the spoon? The pan? The pot?* - Remember - sugar is hot.

Strawberries when long summer days shrink school gates. The farm with her name, we wanted her to claim the straw rows of plump-blue skylark-high as her own. Lifting leaves. *Mine is the biggest! Mine tastes best!* Half in the baskets, half in our mouths. We never wondered what she pondered as bent, intent, diligent she searched and plucked. Maybe it was enough to feel June's warm touch. No matter how many we ate, how sick, how often we had to stop the car to squat and strawberry-pee, watching Wimbledon later, we would still scoff glass bowls of red-sugar-cream, we three.

Mabon damsons hanging heavy at twilight while the river sludged into the estuary; damp-dangerous sinking-mud, *don't go too near*! As the sun sank into the glowering power station, we made castles in Kent chalk and she reached, picked, then stopped to watch the waders walk on water as she smoked – tip-butt-tip.

Blackberries under darkening skies, bush, briar and Tupperware. Fight thorn scratch and stab, wrestle shoulders first, tangling hair into the thicket, we had to stamp brambles flat with wellied feet, forcing scored hands through spider-nets, bluebottle, moth, drowsy wasp, dying bees.

Sticky. She washed. Panned. Boiled. Preserved.

Always sweet. Always three. Always time.

Until there wasn't –

...and her house was empty and we did what we could not imagine doing – we cleared, archived and gave her away.

That day, we two, orphaned, sitting on her cold lino floor, my sister took the jam pan and I the filled jars. *Bin them, you can't eat them*, my sister said, *there's not even a date on them, the labels are blurred.*

And that was nearly thirty years ago and yes, these are the jars and yes, those are the labels and no, I don't want to look too closely because I do not want to find them decomposed. I want to hold them preserved and whole, the tin lids tightly closed.

HIRAETH

We grew up with it,
her bone-deep longing for the hills and roots of home;
her bequest to us
– How long we long for her

HERSTORIES

My mum told me she had a number on a cardboard tag, tied with string to her gas mask, diagonal in a hard brown box across her shoulder.

My mum told me the farmer and his wife were kind. How she loved to look out from the farm up the hill and watch the weather coming in over the Aberdyfi sea.

My mum told me how, evacuated from *Birkenhead (not Liverpool)* at five, she would follow behind the horse-drawn plough on her knees, placing potatoes in furrows carefully, her small hands muddied. How the farmer pointed as the sheep dog bit out the tatws each by each.

My mum told me how, on that hill farm that was home for a while, she fell, one leg down the open drain in the centre of the yard, *I could have drowned.*

My mum told me about her favourite dress, how she loved it, the broderie-anglaise, the white, cotton collar, pink roses around the hem, gold buckle on the belt, how she outgrew it, how it had to be handed down to a younger girl on the farm.

My mum told me how the farmer's wife took her to the village school where she met her brother and her sister (billeted to other families) for the first time.

My mum had photos with friends in milking whites, hair in nets, wellies, coveralls, at Nantwich Agricultural College, laughing up trees, legs hanging through windows, picnics, chasing, lounging. Photos of her in taffeta, lipstick, glowing on the lap of a youth with curls in a dinner jacket, her head slightly tilted, his lips moving towards her throat. Another of him standing by a barn, long legs, slim, black labrador by his side, mug in hand, easy smile. Mum told me how grandma (who ate extra-strong mints), had stopped their marriage because a farmer's life wasn't good enough for her daughter (the daughter she evacuated twice and then brought home only to keep house).

My mum told me she wished she had been a farmer's wife.

My mum thought *maybe* the old farmer was *still* alive, so I drove her back to the pastel sea town, where we asked the baker, where we asked the butcher who said, *over there, the blue house, up three white stairs.* He answered the door, said mum's name, said his wife had died, joined her at the pub across the road. I bought them drinks. A *happy time.*

DEFINITION

Ferme – 13th century, from Old French – *a* **lease or rent**.

Firma – Medieval Latin – a **fixed payment**.

Farming – The cultivation of land for the **production** of agricultural crops, the raising of poultry, the **production** of eggs, milk, fruit horticultural crops, grazing or the **production** of livestock. Includes the **production** of timber, forest **products**, nursery **products**, or sod for **man's use** and **disposal** by **market**ing or otherwise.

Farming – food, feed, factory modelled, intensive, monoculture, fast-growth, high-yield, denude, extract, use, process, monetise, capital, bank, loan, manufacture, profit, maximise, exploit, bankruptcy, suicide, economics, standardize, globalize, resource, productise, ROI, efficiency, supply chain, technology, machinery, toxic, methane, nitrous oxide, pesticides, herbicides, steroids, anti-biotics, E. coli, BSE, Foot and Mouth, TB, swine flu, avian flu, zoonotic, pandemic, cages, contaminants, fertiliser, sewage, run-off, algal-bloom, eutrophication, acidification, ocean bioaccumulation, rainforest annihilation, lead, mercury, cadmium, diethylhexyl phthalate, water scarcity, lost biodiversity, deforestation, market forces, supermarkets, customer, consumer, buy-one-get-one-free, food waste, obesity, diabetes, heart disease.

Farming – A way of / [*for?*] life...

Farming – Food, feeding, evolving, diversifying, hydroponics, aquaculture, fungiculture, permaculture, silviculture, poly-culture, agroforestry, re-wilding, free-range, grass-fed, organic, regenerative, circular, enteric fermentation, carbon sequestration, emissions mitigation, habitat restoration, net-zero, pollinators, companion planting, multi-function, anaerobic digesters, biogas, biodynamic, bioregional, crop rotation, cover crop, no-dig, mulch, compost, swales, small scale, localised, seasonal, co-operatives, potlatch, citizen-science, medicinal, plant-guilds, nutritional density, grey-water irrigation, chicken tractors, humanure, top-soil, solar greenhouses, rain-water harvesting, plant-based protein, eco-systems, community farms, school farms, edible gardens, guerilla gardens, plant pots, allotments, camping, glamping, pony trekking, fishing, WorkAways, Farmstays, Woofing, retreats, education, connection, health.

Farming – Noun / present continuous verb – i.e., the gerund – i.e. – [*we hope*] it keeps on going on and on

I AM NOT SEAMUS HEANEY

- After Churning Day

I keep trying to write about milking, but Heaney, with all his coarse-grained rough-cast keeps trying to have his say. I'm going to tell you about it my way.

You see, the thing that mattered most to me was not the obscene udders, the velvet teats, nor the hissing and pissing as her fingers eased white slip to tin. It wasn't the hot brewery of gland and cud, the flabby milk with flecks of gold, nor the blue-cream striped jugs. It wasn't the man from the Milk Marketing Board, his dray collecting churns, his towering, feathered horse I loved. It wasn't even how much I hated the warm, pulsing, sour-breathed froth which made my cornflakes flop.

What mattered were feet around the docket-strewn table, boots paired on the flagstone floor, Rayburn-warmed. Aunt flicking orders with a sharp whip of towel and tongue. Uncle demanding kippers with his Farming Times, breaking us awake with his shrill whistle, through finger and thumb. *Eat that!* I chewed, gagged on excised bacon fat. He winked his blue-weathered eye, wiggled his nose and ears, told us the cattle taught him how. Mum rolled her eyes, shook her head, smiled. Cousins, all testosterone, jut and knuckles, gripped our weak knees until we squirmed off our chairs to the cold slate floor. Splashing crocks of fetid milk into mugs, they mixed iron tea brick. *Over there!* They shouted and when we turned, they burnt our small sister hands with scalding spoons. Still would, given half a chance, I never learn, they still think I'm slow.

Seamus, it isn't the blood-warm milk I miss, but the coagulated sunlight of kin and kith, heavy and rich.

DEAD THINGS

1

Curly C(uh) is for Cat.
D(uh) is for Dog.

I did not see them drown the baby cats
in the big stone trough
by the small wooden gate
in the place where the bins and boots
sat frozen under corrugated plastic.

I did not see the farm dog crushed –
too loyal, too close.
I did not see my tough calloused
cousin fall to his knees.

2

My uncle said he shot them.
I don't know how.

I was young and scared
of his shouting
and his cattle stick,
so didn't ask.

But the moles
were pads of velvet –
irresistible.
Like sinking fingers
into warm chocolate.

I knew I wasn't supposed to touch –
death was dirty,

but still I smoothed
their soft backs
against my wet cheeks

then buried them home
in dark-crumbled earth –
secret, deep.

CATTLE

1

Goddaughter -

you won't remember your legs around my hip,
your brother's hot hand in mine
as we leant against the metal stall
to watch the vet and your dad
at the wonder of bovine birth.

Fresh from India
hennaed in flowing skirts
I practiced motherhood
- expected beauty, love and peace on earth.

But blood.
But shit.

But heifer lying gasping on her side,
hands stuffed deep into her dark
slimed-feet tied
boots against haunches, cursing -
rope taut, hooves poking black -
then leg, then hip,
then flop of flaccid flesh onto straw.

But it was none of this that made me pass you both
back to your mother's hip
        But twisted heifer head
                her eyes turned white inside
                her silent mouth gaping froth.

I made it to the toilet before I puked
sinking to my knees
shaking sweat

Your mother made me tea.
The cow recovered quicker

New life – black and glossy
sucked greedily.

They still tease me.

2

Cousins –

Do you remember as boys
how you dodged the punch below the belt,
how un-boxed, mis-tackled or grappled
you would double and clutch your crutch?

How then, did you twist tails –
clamp bullocks in the crush –
take your knife –
slash soft skin sacks –
deafen to roarthrash –
slice and screw slimed balls
to circling dogs?

3

Sister –

Do you remember the horns?
Do you remember the heads clamped in metal?
The strong hands racing hot wire
through keratin and marrow?
Do you remember the red bang and bellow?
How some were boiled, cooled –
scooped-goo – a mockery of gelato?
Do you remember how we sanded them,
flower-varnished them,
imagined them mead horns?

SCAR

See this scar?
This fine white line,
a sickle-arc from little finger
across the bridge of raised blue vein
towards my middle knuckle
bending towards bone?

This scar bled
when skin was white and tender-plump –
before it roughed
with sun and thorn.

It bled and I dared not tell
because I scored it climbing the barbed wire fence
where the old sow suckled grunting
and I was not allowed.

It didn't hurt.
My hair in a Purdy cut
I blotted blood on my dark blue jeans –
it blended in with other muck.

The bleeding didn't stop
me creep past stall, over wire
duck the tree and stay low past the cattle sheds
over the metal gate and into the five-acre field.

Five acres of free.
Free from their divorce
from poverty
from *greasy-spic-wop-diego-hairy-legs-lanky.*

I ran the field in my loosestrife top
– blood dripping

free

BETWEEN THE LINES

The Lost Welsh Village   Our last family holiday

(age 10)   *(age 57)*

It was a little valley,   little
With hills on either side,   side
With lots of heather and bracken,
Where the white sheep hide.   hide

I saw that lonely grey mine,   lonely
With houses made of stone.   stone
Just think of all those people
Who lived there all alone.   alone

Just below the quarry,   low
Is a lonely seaside shore,   lonely
The sand is not white, but black,
With stones spread out all over.   stones
                                  spread
                     out
            all
over

WHET

Remember mum?
Waen's field? That soft June evening?
We played Frisbee
when the air was alive with pollen
and we didn't know.

After you died, I walked and cried
beating feet against gravel and grass
to outwalk the black hole of your gone.

Later – in that same field by the river
                    I saw a sheep drowning
wanted to save it –
stripped not quite naked
motherless
        childless
                careless

a foolish girl in knickers slips
down the bank, pulled low
heels – calves – knees – thighs sink
seeking roots in mud
head twists, torso torques, she writhes
flaps her hands against the slow clutch of sediment
        while the sheep floats on
                front legs pedal,
                        head arched skyward,
                              nostrils grasp oxygen
Trapped

she watches as ears, then eyes
then two black holes sink below

      alone

eyes flood heavy
she wants to sink in wet
     let go

She fears the void

Tears scratch white tracks
– a dirt map

gulping deeper
she flings herself forward

arms splay
cheek slaps cool splotch
fingers grasp for what

primeval she lizards
clawing her belly long
her bra fills with sludge
     she pulls
          lugs

                    births herself

                    motherless

JANUARY IN WAEN FIELD AFTER YOU DIED

                                                  a bruised sky

etched

                    with fine-branched-filigree

                                            a tessellation of wings

        swirling black sparks

NOVEMBER WALK AT LAKE VYRNWY

Chrome trees bend
crack-backed
slice-slashed
cloven ash.

Buzzard hangs
sabre-hooked
screeches high
starlings soot.

Sheep bow deep
hidden low
darkened roots
drizzle-blown.

River flies
bronze froth drop
stone-slosh slap
drunken slop.

Red scarf wraps
ice-pinched cheeks
wind-stung eyes
frozen weep.

2001

I couldn't visit when the farm was closed for Foot and Mouth, so did not drive through disinfectant, didn't feel the pull of my child banished to school, not allowed home. I did not see the ministry decree the slaughter of all those cattle, calves, cows, some hand-reared, all known, all housed. I did not hear my cousin lose his bid to have the vet humanely dispatch them quiet. Was not there to witness the army come with guns to scatter bullets like acorns into pens and troughs. Did not hear the bellow, see the stamp, the kick, the crush. I wasn't there to watch corpses scraped, black-tyred to mass pyres. I did not smell the burn of flesh and horn, or inhale the black shroud of fatty ash. I did not draw every curtain, seal every window, turn the TV up loud so the youngest child would not know the horror incinerating in the yard. I did not clean tallow from handles and sills for weeks, not days. I had to stay away. We didn't talk too much about it. There was too much to say.

ECCLESIASTES 3:1-8

Dearly beloveds –
We are gathered here today to mourn the passing of:
Two twisted caterpillar tracks
Three bags of grey hardcore
One concrete pipe, big enough
to hide a child,
Two flatbed trailers
paint peeling, rust rising
    A time to lose
One horsebox, doors swinging
One tipper truck piled with rocks
    A time to throw away stones
Three trailers
One spool of wire
Two huge tractor tyres
One small sheep wagon
One round cattle feeder
    A time to tear down
Seven splintered fence posts
One split water tank
One small red digger
One jammed cattle crush
Three lengths of yellow hose
One crusted muck spreader
    For everything there is a season
One empty bank account
    A time to breakdown
One high barn roof
One taut white noose
    A time to gather stones
One church door shuts
    A time to weep
Another family farm gone
    A time of silence
A time to speak

NOW AND THEN

junkyard/pileofcrushedcars/fishingpool/metalshed/executiveflatsstacked

                                                      a farm
                                       whitewashed walls

                             cattle stalls
                         milking shed

               arcstone bridge
             splashing brook

          blossom orchard
        dappled hens

     smooth brown eggs
     home-baked bread

    elbows creak
      glasses gulp
      cutlery chews

          humming mumble
             ear tags
               price per pound
              yield
                 the ministry

                               family

junkyardpileofcrushedcarsfishingpoolmetalshedexecutiveflatsstacked

QUIETUS

and suddenly the sky turns black
wind   pounces
flings   me flat
grass   flails
to   twist
dust   turns mud
whip   lash
bright   flash
oak   cracks
waist-wide   branch
             smash
cattle   scatter
         trapped
they   turn their backs
against   the thrash
jostle
snort
stamp
hoof-crash
And I think   *they could trample me.*
Then   *are my kids safe?*
Then I see the house lifted like Dorothy's
Then flickering disasters on distant screens
and I think   *Oh this*
                            this is it
                            the thing that is impossible
                            but is possible
                            and could end everything

ROOTING

[An older mother arrives with her newborn; it is clear she loves him fiercely. She is presenting her first born to her aunt and uncle in lieu of her recently dead mother whom she misses intensely.]

She steps across the dusty yard. It is July and the sun is relentless. The dog is chained; kittens weave sticky-eyed around her feet. The air smells of shit and corn. Swallows swoop. She sees her aunt peer and wave through the window. She knows the door will be open but is always unprepared for the heat of the stove.]

AUNT: [*Sitting in her habitual place, by the furnace of a Rayburn, on a low wooden stool as sturdy and worn as she is. She wipes her hands down her silage-stained legs and reaches for the baby with authority, gesturing that the mother sit]:* I used to put mine in the pram down the orchard when they were crying. [S*he pulls faces at the baby, sweating slightly, clearly delighted to be dandling an infant again*]. When my hands were mucky, [*now bouncing the baby on her broad knee*] I used to pop my teat over the side for them to suck. [*She looks at her husband of over forty years, who nods in assent and rolls his eyes conspiratorially towards the mother who watches her son, anxious at her aunt's vigorous jiggling*].

[A word about the aunt. She is a farmer's daughter and a farmer's wife who has raised four children and has lost count of the number of grand and great-grandchildren. She rules the house and the yard with a sharp tongue and if needed a stick to the flank or a clip around the back of the head. She can milk, bake, pluck, gut, sling a bale and drive a tractor. She adores her husband even though they fight loudly, fiercely, often.]

UNCLE: *[Looking up from lists of numbers on cards. To the mother]:* Eat something woman, there's not enough flesh on your bones for the milk. [*It is an order, not a suggestion. He is used to people doing what he says.*]

[The aunt thrusts the pink baby back to the mother, lifts the already steaming kettle back onto the stove and walks around the long, paper-strewn table to the larder door. The catch rattles and she treads down three worn stairs into the cool slates of hanging hams to reach for the moist, jam-filled, icing-dusted cake she has somehow found time to bake between feeding and mucking out. Slabs are cut. Tea is poured; copper, hot. The uncle watches as the mother eats. He wiggles his nose at the baby, then his ears. She has never known how he does it.]

UNCLE: I learned it from the cows.

[A word about the uncle, her mother's older brother. Evacuated in the war to a farm, he fell in love with the land and then, on return, with his wife. They have had smaller farms before, but this is the land they will die on. He has had knees and hips replaced from long hours on cold floors. He is known in the cattle markets as man of quick wit and sharp mind. His handshake is strong. His word, his bond. He is successful but when out of his wellies, cap and overalls, has none of the usual indicators of wealth. Impatient, demanding, critical, he speaks his mind and does not suffer fools. He loves his wife.]

[The baby, smelling milk, cries and nuzzles.]

UNCLE: [An order. Kind but firm]. Get your teat out then, he needs feeding.

[The mother does. The baby feeds, gulping. The aunt and uncle watch closely; used to checking latching on and milk flow. Awkward, embarrassed, the mother gets up to move as she feeds].

UNCLE: Sit down while you're milking that baby! [*He gestures the seat].* You don't see heifers walking round while they feed.

MILK'S GOTTA LOTTA BOTTLE

Where does milk come from?
It comes from cows.
How do cows make milk?
They have to have a baby-calf.
Don't the baby-calves need milk?
Yes, they do, but they are taken away so that you and me can drink it with tea.
Are the calf and the mummy sad?
Yes, very sad. They call for each other. Loudly.
What happens to the calves?
Some are killed, some are fattened for beef and some have calves of their own.
So how many times do mummies have to give away their babies?
Three, four, five times, depending on how much milk they make.
What happens when the mummy cows are too tired to make calves and milk anymore?
They become dog food, burgers, processed food, you know, the kind of things we eat.
-
Do you know one more thing?
Do you know that when
the mothers are crowded into pens
waiting for their end,
do you know that
their teats leak milk?
Do you know that milk pours
from their sore
swollen udders
on to the shit-soaked,
knee-deep straw?
Do you know
that still they call?
Pour milk and call.

TIPPING POINT

Dog lead, handle creak – I
sneak free. *Take me!* Frown, tut,
strap pram, tug lead, push, stomp, steam.

White van, narrow verge, dog pulls, you tip –
nettle ditch, white welts – I blanch, you scream.

SHIPWRECKED

I find her heft-rolled onto her broad, unsheared, pregnant back
                          legs skyward
face black with blood where crows peck her live eyes

She rasps, gasps, grasps at life, suffocated by weight and child

approach   slowly   bow   low   murmur   intone
in mud I kneel
fingers find pulse in fleece

press gentle
palm calms us both

I breathe then push
    she lists

pause to lever knee under fleece

then push again
       until

she tilts, tips, rolls like a bulging burlap sack
spilling onto her side

pant   acclimatise

knees – slip – knees

struggle to find feet as land sways
lean in warm – can't capsize again

foot by foot
     haltingly we rise
        squint
           blink
              shake
stagger into our day

FORESHADOWING

As I walk from the woods, down the hill to the pebbled sea, I see a woman on a horse riding ahead of me. I watch her easy trot, then stop, holding the kelpie's stamp-dance between her knees.

Bareback, only loose reins and poise bonding her between wither and croup, she rides to where our paths will cross, by the gate with the early blackberry crop. She and bay are damp with sweat-spray, her cobalt smile reflects my gaze.

They have been along the paths behind the sea, *checking the gates* before she drives her highland cattle to graze the heath. Her hands are strong, her look direct. *Not bad for a job, is it?* she says. I follow with my eyes as she rides away.

Centaurine, they gather a canter across the gorse, lift light and dust, flying-woman-horse.

I want my skin to be brown-bared, my hair tangled-breeze, I yearn to ride unsaddled, spray on my skin, between land and sea.

LIMINAL

*

So I ask Liz
and she says,
Yes
she can teach me
bareback,
*Barefoot too
why not?*

we laugh

dream
of riding
the shoreline
naked Godivas
at dusk

*

*Pelvic floor in   up
Vertebrae stack
Balance your seat bones around her spine
Align sternum to pubic bone
Sink your weight
Rest elbows to ribs
Tuck in chin
Drop shoulders
Let legs drip down her flanks like ice-cream
Pause   focus
See the serpentine
Steady   exhale
Walk   trot
Halt*

Canter this time
Let your hips move
Keep your spine in line
Let her breath
guide your breath

The arena demands
more than the dream.

*

We never rode the line of sea
and land that winter –
sickness reared –
hooves crashed hard on soft.

*

Now silvered
we wind new charms
by moonlight
skin-to-skin
and tempest-tossed

DRIVING TO WORK

I pass yearlings
grazing mallow
on the Meifod verge

I see Perygl! Danger!
hooves near tarmac
and smack of tyre

I pull up,
but hesitate –
gate my urge to
herd them safe –

I understand
their longing for
abundant,
if forbidden land.

FERAL

(For Clare)

speckled thrush calls us
from motherhood to creep
barefoot along paths of leaf-rot
twig-snap mud-ooze bramble-snag
– untangled we whisper wilds

WITCH

which
witch

      weird
      bitch

           stone her
           drown her
           in a ditch

Now

We
Granddaughters
of those who
Burned   and   Drowned

      Gather cindered Knowing
      from the damp-dark-fertile Ground

CRAFT

blackthorn stabs sharp spikes
flings ivory confetti
petals hiding pain

TRESPASSERS W

I'm in the piano-tent, near a stream, on farm fields, up a hill somewhere in Wales. Fiddles and banjos play up the dirt track, through mirror-hung trees, round the pop-up coffee shack. A grass amphitheatre embraces dreadlocks and pixie fringes, bikinis and porkpie hats, babes in slings and aging rainbows smoking spliffs. It is unnaturally hot for the end of May, people sweat, I sit in the shade.

I am part of the earnest eco crowd, more or less sober, a mixed mob. We have listened to the Land Workers Alliance, hemp growers, ethical cacao, the man with piercing blue eyes who sailed from Scotland to Denmark, in November's spewing waves, to learn about his roots and wood.

I am still here as the sun-notes rise and amplify, more through hero-worship than commitment to Right to Roam, but if Jay Griffiths is there, so am I; I have stalked her through books since I found her in *Wild*. She doesn't disappoint, there is passion and mischief in her words, she shapeshifts the page, dances with etymology, mercurially knots ideas into intricate patterns which infiltrate, dissipate, dissolve the taken-for-granted world.

She and Sam Lee, folk singer and old-song-conservationist, move us with story and melody, stoke our desire for fenceless wandering. I nod as Jay remembers the summers of our childhood; clouds of moths spiralling around street lamps, night driving made dizzy with snow-storms of headlit wings, sticky coated windscreens with corpses even on the shortest trip. I nod too in recognition of the quiet perfection of the wild-morning-swim, her in a lake across a farmer's field, me in the river at the foot of my garden.

*

Elsewhere, elsetime, I horseback the sunset-solstice-hour with the landowner of another Welsh hill. When I broach the right to roam, she shudders and tells tales of walkers with dogs off leads, scaring horses, chasing pregnant sheep. So much land fenced in, no right of way. I slap a horse-fly bite.

And yet, and yet, when I walk the further river bank, where people jump the gate to fish or splash the shingle river-beach, it pisses me off to pick the litter I didn't leave; the plastic bottles, empty Pringles tubes, sausage-roll wrappers, cans of warm-dreg-slop-beer.

And yet, and yet, this huge-wide view of skylark-heather where we ride, this screech of buzzard, this white-hop of rabbit scoot, surely this should, this should, this ought to be available for all to see, hear, breathe. For humans are not so good at protecting what we do not love, let alone, what we do not know, and if ever there was a time to love and protect our earth, it is now and now and now.

*

Frost says; *Good fences make good neighbours.* But he is not convinced, *Before I built a wall I'd ask to know / What I was walling in or walling out, / And to whom I was like to give offense.* Raised on *Lassie* and *Black Beauty*, my gypsy heart yearns for the freedom of the horse-drawn caravan and a life of simplicity; I want to roam, but don't want people to roam too close to me.

In - out / mine - yours / included - excluded / contained - free / domestic - wild / vagabond / vagrant / homeless / landless / settled / safe / home.

*

I sit nodding with Jay in the stippled sun, buoyed by words, drenched in lyricality, churning with hypocrisy because I wouldn't like just anybody, the uninvited, the loud and litter dropping, the stranger, to enter my garden without permission to dip into the river to swim.

Jay folds her papers, Sam Lee ducks a bow. I leave the crowds, the grounds, and walk the road away from the laughter and notes until I step over a gap in someone else's fence, hush past lizards basking on warm stones, tread softly to a stream in land I do not own and dabble my toes.

FIELD

©

© An area of land, used for growing crops or keeping animals: surrounded by a fence or hedge | to keep in | out
© An area, covered with grass for playing sports: the school | sports | football | hockey | rugby | field
© An area of activity or interest: the field of history | science | medicine
© Competitors taking part in a race | interview | activity: *We have a strong field of candidates*
© To catch or pick up the ball after it has been hit in a game such as cricket | baseball | to try to prevent the other team from scoring
© To deal with a question | often by not answering it directly: *He fielded some awkward questions very skilfully.*
© To produce a team of people to take part in an activity | event: *The company fielded a group of experts to take part in the conference.*
© A division of a database: a collection of similar information on a computer | names | numbers

Where is the common? The free? The green? The easy? The soft succulent smell of sweet soil? The dabble of brook? The lizard? Slow-worm? Toad? Where the trunk of shade? Where the cowslip? The primrose? Harebell? Columbine? Where the rabbit? The deer? The hare? Where the skylark? The buzzard? The swallow? Where the bee? Cranefly? Butterfly? Where the wind, the sun, drizzle, mist? Where the dawn, the dusk? Where will we make love? Bring our children? Dally? Chase? Picnic? Read? Daydream? Rest? Breathe freely? Doze?
Where?

SHIFTING BASELINE SYNDROME[1]

uncut verges wave –
red bug splat smeared windscreen flat
heralding new life?

"Shifting baseline syndrome'?	Coined by marine biologist Dr Daniel Pauly in 1995 (Ecology Training UK)
What?	'A gradual change in the accepted norms .... of the natural environment'
Why?	Without memory, knowledge, or experience of past environmental conditions, current generations cannot perceive how much their environment has changed because they are comparing it to their own 'normal' baseline and not to historical baselines.' (https://earth.org/shifting-baseline-syndrome/)
What?	Do you remember?

CONFLUENCE

So Liz and I are chatting by her car −
(and if you knew me and she
you would know that we do)
 − leaving words of love and lettuce

  and then
    and then
      Pterodactyl beats
        the air above our heads
          gold beak arrows Cain
            stick toes trail Tanat
              a crucifix at Vyrnwy
            we flinch to stoop
            are whooshed to shush
            under its low grey canopy
          still we stand naked
        river smeared
      dripping
    silent
  O

WHERE WAS NOAH WHEN THEY NEEDED HIM?

It rained
in the night.
There are worm
tracks; etchings
like neurons
stretching for
synaptic cleft.

I tread slow
through
lattice-fret,
intestine-coiled,
pink flesh,
fearful of
smearing death.

A friend says
they rise
from flooded
burrows
to survive.

Then why, I ask,
are they all dead?
Life, she says,
doesn't always
make sense.

GARDENING

I share my garden –
leave half wild.

Bats nest, pheasants forage,
frogs hide in the damp shade of weeds
under the weeping birch tree.

Yet still I mowed a slow-worm
skinning silver scales to glistening pink.

It is not true that they carry on as two.

*

                                          Rock a bye baby on the tree top

Hearing shots
I crane to see
my neighbour shooting
nesting
resting
rooks from trees

                                 When the wind blows the cradle will rock

I scroll a retreat,
find rooks are *Red –
Endangered*
...killing *permitted*
when home is *threatened*

                                                When the bough breaks

I watch them circle confused,
squawk-flap
crowd-throng

we/they/you
did not belong

                                                      the cradle will fall

I throw
cowardly corn
onto my lawn

and down come our babies cradles and all

HARVEST

Wicker basket brims.

Damsons shine purple-plump.

Sweet juice drips, slicks silk skin.

Rooted in our collie dog;

his death a blossoming.

CARE

(For Phil)

You could have waited
for the lengthening May light
to tame
the shining ivy, lacy
hydrangeas, honeyed jasmine.

Instead, Imbolc numbed
your deft wood-stained fingers as
you sawed, hammered, nailed,
tenderly tugged
tendrils to hide nesting boxes

So now, at Beltane,
as I hang washing in the
crisp spring light,
sharp pin-point calls bring bluetit parents,
mouths full of worms, for your kith

TOO LATE

You turned up late. You had something to do. We all do.

You turned up late so I couldn't show you the track I walked in white blossom May, with strangers, as the sun began to set, to a place where benches squatted low in mud, where fire burned in circles of iron, where me and fifty others, sipped nettle tea or beer and listened to songs, strings, eventide carolling.

I wanted you to walk with me so I could tell you about the anthropologist who recorded nature's sounds for posterity, as remedy, about the man in wool with the old leather hat hiding a flushed younger face who gathers melodies from the old ways, preserving them against decay.

I wanted to tell you how in Kent, as a girl, playing out with friends; how this bird was just part of the hides, the seeks, the shouts, the squeals. How as the other birds quietened, this one kept us company as we finally followed the purpling-sky-star-rise home – no phones.

I wanted to tell you how when a hundred feet snaked in a line elephantine to the foot of a tree, how when fifty mouths sealed and a hundred ears strained, how I sunk to my knees when I heard those old notes again.

How then there was

[   ]

I wanted to tell you how I wanted to be alone, in the black, the shadow, the silence, the empty, how I wanted to lie crying in the mud – smeared, streaked.

I wanted to tell you how then fingers moved across strings, raising octaves, stretching staves, reaching

Then I wanted to tell you how after time,   long time,   so much time,   more time,   most time,

    three - low - clear - notes blew from the high night branches

    How the strings replied.

How across the night, the pair played, jammed, riffed, gigged, gathered harmonies on staves of leaves.

I wanted to tell you it was a holy moment dripping soft warm gold.

I wanted to say to you, this, here, this, hear, is where I find my faith, this is my cathedral, this is how I pray.

I wanted to tell you how, when the group walked back to the cars and the roads, I lingered with a few, how in that wood the bird sang a

        single

            aria

          alone

How I did not want to leave. How the song was so loud, so sharp, so clear. So strong. How familiar it felt to me, but how in hearing it, I realised I hadn't heard it for so long.

I wanted to say, come, come, come my children, sit in silence with me in trees and hear this bright song, pass the remembering on, let it be your children's children's song.

YOU DON'T KNOW WHAT YOU'VE GOT TIL ITS GONE

After Joni Mitchell.

For Sam Lee and the Nest Collective

Have you heard the joke about the anthropologist, the feminist and the teacher

?

How each paid to sit on the earth, in a wood to hear a lone nightingale sing.[2]

---

[2] *Silentium non est aureum*

In the 1960's, during my childhood, over 70,000 pairs of nightingales nested in the UK.

In 2024 there are fewer than 5,000 nesting pairs.

Causes: lost thickets and scrubland, over population of muntjac deer, industrial farming, building, fossil fuels, herbicides, pesticides, artificial fertilisers, climate change.

In 2023 in this wood in Gloucester there were six calling, mating pairs.

In 2024 in that wood, on that damp and wood-smoked May night, only one lone bird sang.

The rest

was silent.

ROAD-KILL

Driving home,
daydreaming over hills
and someone else –
I hit a pheasant.

Head wedged in radiator grill
butt hanging out
wings splayed –
trying to push away.

My neighbour
hung, plucked
baked it
       – murder made palatable.

*

Roadside –

tawny owl
head
the size of hand over fist

bodyless
its big-black-wide-open-eyes held mine.

      Lot's wife
      I dare not turn aside.

*

In my rear-view mirror, I saw it –
hips crushed,
front paws scrabbling tarmac.

Maybe I should learn to use a knife
    – slit throats
        let suffering out.

MAN VERSUS GREAT BIRNAM WOOD

Here, upon this bank and shoal of time

a corrigendum will be printed tomorrow, and tomorrow, and tomorrow – it creeps in at petty pace, but does not resolve what has happened today.

As a result of yesterday's (lack of) negotiations (or information), (or consultation), (or mitigation), there was a corrigendum in which the word 'tree' was deleted for the last syllable of recorded time, then was heard no more.

A corrigendum (signifying nothing) will be issued in due course to address the issue of (your) *need to tidy up around the car park area.*

An erratum notice is published to correct errors in policy that were inadvertently created by an idiot, full of sound and fury, felling the way to deep damnation and a dusty death.

Something to do with reducing bottlenecks in the flow of the river to allow water to run off quicker and therefore reduce the risks of flooding, Something to do with the stability of the bank – maybe a hedge would stabilise it more than taller
trees
that could
topple,
taking
the bank with it,

Are tales signifying nothing.

Their main concern (vaulting ambition tainted with fear) *is the impact on the limited travel infrastructure in this area.*

Here comes a chopper to chop off your head!
Here comes a candle to light you to bed.

Out, out, brief candle!

They should have died hereafter.

Vulpes	VULPES VULPES	Vulpes
suddenly	Top of the woodland food chain: a diet of birds and beetles, rabbits and rats.	between the arc of trees
in the grey		damp heat
of summer	In 2013, **96 per cent** of Welsh farmers said that fox predation of lambs had an impact of **£9.4 million** on their income	dawn
a shadow		trots.
With green		eyes it fixes me,
haunches	The benefit to farmers of foxes killing rabbits, thereby protecting crops, is estimated to be **£7-9m.**	me low,
ears twitch		me still,
nose slows	Reducing fox numbers by **43 per cent** resulted in a **three-fold** increase in breeding success for lapwings, golden plovers, curlews, red grouse and meadow pipits [all threatened populations]	my breath.
I bow,		hunker humbly
watch it		move – track, trunk,
glade where		russet roots uncurl
as mother.	Hare densities at a farm in Leicestershire **declined** when predator control was carried out [hares are protected in Scotland, not England or Wales]	
Chew-chase,		tug-tussle,
moss,		bracken,
the dyad		dryads roll-spin-play.
Please stay.	It is **illegal** to hunt foxes with a pack of dogs [people around here still do]	
Clumsy,	Free-running snares such as a wire loop, do not relax when the fox stops pulling [a slow death from injury, starvation or strangulation]	I unravel,
break my		silent vow
of		smallness.
Vixen turns,	When inspecting snares, it is essential that a means of humane destruction of a snared animal is available. A .22 firearm or a shotgun is suitable for the purpose.	freezes,
barks –		
they dive as		shadows
into the	Air weapons should not be used, as they are not sufficiently powerful	dingle-dark

MINK

Performing at brook-ford:
floodlit by sunlight,
a rich shimmy of ebony
zig-zags and darts
the bridge catwalk.
Black velvet magnificence
undulant radiance
muscular brilliance
drum roll, applause.

*

Slung, dangling
from shoulders of farmers

high-waisted in waders
through damp morning fields.

Accused of incursion
non-native invasion

attrition of voles and
ground-nesting birds.

Bunched by their hindlegs
the wilted cadavers

droop their keen claws
towards river and dirt.

*

Denmark in cages
  dread covid mutation
    libations
      of corpses
        spill
          on soil
              like oil

HAWK

...I've brushed my hair, tied my shoes, am leaning in to shout *goodbye*
when there, through the window, I see you
   brown, gold, white-striped tail high
       alight
           yellow eyes
                slice-rusted bikes, firewood piles
                incise tangled ivy

                    wasps buzz drowsy

                        sparrow-squabble
                            tremble
                                muffled
                                    falls

                                        frozen in frame

I stand behind glass –
a caged shadow
watching.

Son breaks the spell –
You twist, blink
glide away.

We gather uniform –
tick-tock
on with our days.

SOAR

It is dawn and I walk in leaf-green light
alone. Hawthorn froths white foam, celandine
shines, bluebells droop with dew. Through the soft
sap of trunk and bough, dunnock, thrush, sparrow,
wren, fling riotous notes of accompaniment.

A commotion to my left stills my feet, quiets my
breath. By the brook, a chatter, clatter, screech. I
turn in time to see a tawny owl, cream belly,
chestnut wings spread wide-eyed, pursued by
blackbirds along the rill.

The wild claims me.

*

My teenage son is in bed.

I wonder if he will ever seek sunrise, desire
the power of swoop and wing, or if he will Hansel-lose
his way in the | shiny | binary | of screens.

*

A buzzard shrieks, I tilt my chin, lift to
circle-glide the wind.

FIREFLY

luminous
numinous
phosphorescent
pulse
solo
halo
glow
Oh!

BREAKFAST

There are worse places
to have breakfast
than down by the river

under the old hazel tree,
where the sun rises
through a lattice of branches

greening the steam
from my coffee cup
ambering the frost.

DUCKS

Still under willow
two ducks glide downstream -
easy surrender

ASUNDER

You scooped ducklings
in your safe hands,
one by one
from the white line
on the busy road

coaxed, stroked, whispered
reassurance to them
returned them to the river
and their mother
that last summer you lived at home.

A SUMMER EVENING

Dangling my feet in the cool evening river
I search for swallows, watch gnats cluster low
along the slow tide.

A mayfly floats by, drowning.
I cup my hands and lift it out.
A flimsy thing, the kind that bounces
on warm summer air as if on springs.

At first it seems dead,
but then antennae move; tail; wings.
I rest it on a twig

step gingerly in to find myself
ringed by buoyant corpses; mayflies or freshwater
shrimps drifting in the pollen-coated-current.
I can't save them, too dead, too many.

I feel a tickle, look down to see inch-long, darting
fins nibbling my skin.
I lean forward and swim.

PRESENT

If you sit still and quiet in the same place,
time and time again, you will get to know

robin, chaffinch, blackbird, collared dove,
cleavers, nettle, red campion, blue aconite,
ant, bee, woodlouse, worm, orange-tipped
butterfly, tadpole, toad, rock, moss,
sycamore, blackthorn, willow, ash.

If you are still and quiet often enough, they
will get on with being bird or plant or insect,
right up close to you. Right up close
so you can see their eyes, their feathers, their
petals, their tiny, ordinary, intricate beauty.

HOPE

        mycelial
    conversations
      weave
        roots
    compost
   rot
feed
  seed
    flowers
aflame

MEDITATION WITH A BARN OWL[a] AT TREFLACH REGENERATIVE FARM

There, *aum-ing* meditation, on a soft summer evening,
the old barn flings off cobwebs, opens its doors
basks in the glow of burnished oak floors.

There we exhale beyond wide golden glazing,
to apple-fed piglets, bright shining chard,
free-grazing cattle on shimmering grass.

There in the hot slant of midsummer lustre,
we tilt and turn to the low hum mantra,
whirling a call for light, *Ya Nur*

When later leavened we part,
reborn in radiance –
a hand reaches for arm, calls – *There! There!*

And there across sloping fields of sunwash
it glides – a blaze of white from blue
falling empyrean through sparkling pasture
luminous rays outstretch and swoop

whirling freely, tilting, turning
a beam, it floats unfettered on joy
gleaming-wings wide, ethereal, spacious
embodied light, the sunset's envoi

---

[a]   (British Trust for Ornithology)
Worse for: Cars, agricultural chemicals, pesticide seed dressings, rodenticides
Better for: Efforts of volunteers, erection of nest boxes, hedgerow trees, old farm
            buildings, prey-rich rough grassland
Declined: 1970s and 1980s.
2015   UK Green List
Now: May exceed 10,000 breeding pairs.
Humans: Can make a positive difference

HARES AT 42 ACRES REGENERATIVE FARM

I saw a hare. Or four hares.
Or the same hare four times.

Anyway, it doesn't matter
other than that for four moments,
I stood stilled and quiet
as the long-legged, lithe being
stopped its sprint to look back at me -
yellow coat streaked black like corn
against a dusky sky.

We two shared stares.   Unstartled.   Unhurried.

Then, satisfied at what was sniffed
the hare loped into long grass
to carry on with who knows what
and I watched after it.

NOTES

Definitions – the definitions are taken from a range of sources in particular the Permaculture Design Course I did at Treflach Farm with Steve Jones of Sector 39 in 2021

Between the lines – the first poem really was written at the time of my parent's divorce, on our last family holiday, to the area I now live in. I thought the poem was lost but my dad kept it, and returned it to me for my fiftieth birthday. When I re-read it, I could see what I couldn't articulate then.

Vulpes Vulpes – The words in the centre of the poem are taken from the following sites:

> https://www.discoverwildlife.com/people/do-we-really-need-to-control-foxes-in-the-uk/
>
> https://www.woodlandtrust.org.uk/trees-woods-and-wildlife/animals/mammals/fox/
>
> https://www.gov.wales/sites/default/files/publications/2019-05/rural-foxes-fact-sheet.pdf
>
> https://www.gov.uk/hunting/mammals

THANK YOU

First and foremost, I am so grateful that I live in a safe and abundant land, where there is still soil healthy enough to produce food, rivers which still have fish in them, where the water is safe enough to drink and the air to breathe. My wish is this for all sentient beings, human and more than human. May we learn balance, gratitude and 'enoughness' so that we can live in harmony with the web of life now and for future generations.

Thank you to my mum and dad who planted us firmly in nature and taught us the art of going for a walk, no matter what, no matter where.

Thanks to my sister for all our shared walks, actual and metaphorical.

Thanks to my aunt, uncle and cousins for sharing your farms.

Thank you to local farmers for all your work and for letting me wander your lands.

Thank you to so many wonderful friends who have woven lives, walks, talks and time with me. Including you P.

Thank you, Elizabeth Turton, for everything, especially your earth-deep connection with the wild

Thanks to my first readers, Molly Cullen and Zoe Clarke for your detailed attention, belief and honesty. And to Ceri.

Thank you to Clare Best who was my reader and mentor in the final stages of the collection - always a pleasure.

Thank you, Mark Davidson of Hedgehog Press, for giving my poems a home so beautifully and thank you to Booka Bookshop, Oswestry, the best independent bookshop ever.

Thank you to Hannah White for your photography. Clare Best, Shanta Everington and Sarah Law for support of my writing and your words.

Thank you to Ian and Stef from Treflach Farm, Casha and Barbara from Babbinswood Farm, Tom Adams the Apple Man and Vera Van Heeringen and Jock Tyldesley for joining me in the launch.

And final thanks of course to my sons, Matty and Ben for all the walks, all the shared time, the music, the piss taking, the movies, the shouting and the laughter...love you loads.

ACKNOWLEDGEMENTS

Ecclesiastes 3:1-8 was the winner of the 2024 Bournemouth Poetry Prize and is published in 'Di(still); the competition anthology, Bournemouth University.

2001 was short listed Mslexia Poetry competition spring 2024 and published in Mslexia.

The pamphlet from which this collection has grown was one of 7 finalists in the 2024 Pamphlet competition and one of five finalists in the 2024 New Voices Pamphlet competition both run by Cinnamon Press

Similarly, the pamphlet was short listed by Hedgehog press in spring 2024's Proper Poetry Pamphlet competition

Man Against Great Birnam Wood was one of 7 poetry finalists in the London Independent Story and Poetry Prize

Shipwrecked was in the top 9% of the Bridport Poetry Prize entrants in summer 2024

Vulpes Vulpes was long listed for the Canterbury Poetry Prize in August 2024 and appears in the competition anthology.

January in Waen Field after you died and *Soar* were published in 'Thin and Sacred Places', (Ed Sarah Law), Amythyst Press, Summer, 2024

Foreshadowing, Hares at 42 Acres Regenerative Farm and *Confluence* were published by Hazel Press in January 2024

www.ingramcontent.com/pod-product-compliance
Lightning Source LLC
Chambersburg PA
CBHW021639080526
44584CB00015BA/1613